baby RACCOONS

KIM THOMPSON

CREATIVE EDUCATION • CREATIVE PAPERBACKS

CONT

ENTS

I AM A KIT.

I am a baby raccoon.

I have a mask of dark fur. My tail has stripes.

I was born in a den inside a hollow tree. My mom had four kits.

We drink our mom's milk. At first, I could not stand. I weighed as much as a deck of cards.

I am two months old. I walk in the woods with my family. I swim in streams.

I am a good climber. I like high places.

I use my paws like hands.

My paws pick berries. They catch fish. They open garbage cans.

I am nocturnal. I sleep all day.

SPEAK AND LISTEN

Can you speak like a kit?

Baby raccoons chirp and squeal.

Listen to these sounds:

https://www.youtube.com/watch?v=wX6T3-uuyMU

KIT WORDS

den: a home for mother and baby raccoons

hollow: empty inside; dug out

nocturnal: asleep all day and awake all night

paw: a raccoon's foot

READING CORNER

Chanez, Katie. *Raccoon Cubs in the Wild*. Minneapolis, Minn.: Jump!, Inc., 2023.

Einhorn, Kama. *Raccoon Rescue*. New York: Clarion Books, 2019.

Riggs, Kate. *Raccoons (Amazing Animals)*. Mankato, Minn.: Creative Paperbacks, 2023.

INDEX

PUBLISHED BY CREATIVE EDUCATION AND CREATIVE PAPERBACKS
P.O. Box 227, Mankato, Minnesota 56002
Creative Education and Creative Paperbacks are imprints of The Creative Company
www.thecreativecompany.us

LIBRARY OF CONGRESS CATALOGING-IN-PUBLICATION DATA
Names: Thompson, Kim, 1970- author.
Title: Baby raccoons / Kim Thompson.
Description: Mankato, Minnesota : Creative Education and Creative Paperbacks, [2026] | Series: Starting out | Includes bibliographical references and index. | Audience: Ages 4-7 | Audience: Grades K-1 |
Summary: "Introduce beginning readers to the world of baby raccoons with this life science starter. Includes photos, a labeled animal diagram, "Make a Noise" section, glossary, and further resources"-- Provided by publisher.
Identifiers: LCCN 2024043259 (print) | LCCN 2024043260 (ebook) | ISBN 9798889897538 (library binding) | ISBN 9781682778395 (paperback) | ISBN 9798889897668 (ebook)
Subjects: LCSH: Raccoons--Infancy--Juvenile literature.
Classification: LCC QL737.C26 T465 2026 (print) | LCC QL737.C26 (ebook) | DDC 599.76/321392--dc23/eng/20241209
LC record available at https://lccn.loc.gov/2024043259
LC ebook record available at https://lccn.loc.gov/2024043260

DESIGN AND PRODUCTION
Design by Rhea Magaro
Production by Beeline Media and Design, Inc.
Art direction by Tom Morgan

PHOTOGRAPHS by Alamy Stock Photo/Radius Images, 6-7; Dreamstime/Bsheridan1959, 5, Isselee, cover, 2-3, Svetlana Foote, 4; Shutterstock/Agnieszka Bacal, 10-11, Bohbeh, 13, Biego Puyal Puente, 7, Jay Ondreicka, 9, Orest_U, 8, Robert Eastman, 11, Sonsedska Yuliia, 12, 14

Printed in India